One Melts Away

written by

Naomi Glasarth

illustrated by

Rubén Carral Fajardo

a burning butterfly book

One Melts Away Copyright © 2018 Naomi Glasarth
All rights reserved.

Cover, layout, and illustrations copyright © 2017 by Burning Butterfly Books
Illustrations by Rubén Carral Fajardo and are created digitally.

Library of Congress Control Number: 2018949839

Hardback ISBN 13 978-1-947344-83-9

Published by Burning Butterfly Books
Camp Verde, Arizona
www.burningbutterflybooks.com

*For Susan who loved winter and teaching
and believed in me.*

Words to Know

snowman	sticks
snowmen	slide
happy	drive
stand	sing
line	door
melts	play
away	tree
go	sled
gate	zoo
dance	look
pin	fun
fish	doze
knobbly	sun

Verbs to Find

Verbs are *action words*.
Can you find these action words in the story?

waiting	singing
going	standing
dancing	sledding
fishing	looking
sliding	dozing

10

9
10
8
7
6

ten   nine   eight   seven   six

Ten

5

4

2

1

3

five

four

three

two

one

happy snowmen.

Ten happy snowmen

waiting in a line.

One melts away

and then there are nine.

9

9   nine

8   eight

7   seven

6   six

Nine

5
four
three
two
one

5   4   3   2   1

five   four   three   two   one

happy snowmen.

Nine happy snowmen

going through the gate.

One melts away

and then there are eight.

8

8

7

6

eight

seven

six

Eight

5
five

4
four

3
three

2
two

1
one

happy snowmen.

Eight happy snowmen

dancing on a pin.

One melts away

and then there are seven.

7

7

6

seven

six

Seven

5
four
four

5
4
3
2
1

five   four   three   two   one

happy snowmen.

Seven happy snowmen

fishing with knobbly sticks.

One melts away

and then there are six.

6

6

six

Six

5

4

2

1

3

five

four

three

two

one

happy snowmen.

Six happy snowmen

sliding down the drive.

One melts away

and then there are five.

5

Five

5
four
five

4

2

3

1

three

two

one

happy snowmen.

35

Five happy snowmen

singing at the door.

One melts away

and then there are four.

4

Four

4

2

3

1

four

three

two

one

happy snowmen.

Four happy snowmen

playing around a tree.

One melts away

and then there are three.

3

Three

2

1

3

three

two

one

happy snowmen.

Three happy snowmen

sledding by the zoo.

One melts away

and then there are two.

2

Two

2

1

two

one

happy snowmen.

Two happy snowmen

looking for more fun.

One melts away

and then there is one.

1

One

1

one

happy snowman.

One happy snowman

dozing beneath the sun.

One melts away

and then there are none.

O

Zero

happy snowmen.

Treasure Hunt

Can you find the snowman wearing the...

red vest

purple hat

blue coat

yellow scarf

green boots

pink earmuffs

gray belt

orange mittens

black sunglasses

brown cape

Prepositions

Prepositions are words in a sentence that show a relationship between a noun (or pronoun) and the rest of the sentence.

Which snowmen are these? These are the snowmen standing *in* a line. Those are the snowmen standing *on* a line. The other snowmen are standing *around* a line.

The preposition changes the snowmen's relationship with "a line".

Find the preposition in the story that completes each sentence.

Ten snowmen stand _____ a line.

Nine snowmen go _____ the gate.

Eight snowmen dance _____ a pin.

Seven snowmen fish _____ knobbly sticks.

Six snowmen slide _____ the drive.

Five snowmen sing _____ the door.

Four snowmen play _____ a tree.

Three snowmen sled _____ the zoo.

Two snowmen look _____ more fun.

One snowman dozes _____ the sun.

Naomi Glasarth draws upon her vivid imagination to craft stories for children of all ages. From colorful picture books to middle-grade science fiction and fantasy, she sets her creativity free to run wild and invites you along for the ride. When not crafting her next exciting adventure, she can be found enjoing the sunshine of central Arizona where she imagines seven fantastic things eah morning before breakfast. Her gods, Blue and Bear, contribute to her stories by curling up beside her as she writes. Sometimes they slobber on her leg.

Rubén Carral Fajardo seeks to transport the viewer into the world of his art. As a professional digital artist, he fills his illustrations with color, emotion, and a sense of adventure overlayed with a touch of the mystical.

He lives in Spain.

Hello bright and beautiful butterflies. Thank you for reading *One Melts Away*.

If you enjoyed counting backward and learning about verbs and prepositions from our adorable snowmen, you will love counting forward with our dragon Cici, in *Cici Counts by Ones*.

Please visit us at www.burningbutterflybooks.com where we ignite imaginations one story at a time.